FIGHTING FOR A
GREENER WORLD

By

Edward Kane

INTRODUCTION

"Fighting for a Greener World" brings together the latest science on Climate Change and the most innovative global remedies and solutions. Climate Change is worsening with temperatures and sea levels rising and the ice sheets in the Arctic and Antarctic melting at unprecedented rates. Climate change is spawning extreme weather including months of historic flooding in China, tropical temperatures in Siberia, massive back-to-back hurricanes, unprecedented wildfires and scorching summer temperatures. Global scientists warn that the Earth is on a collision course with Climate disasters unless action is taken now to drastically cut greenhouse gas emissions.

Green Travel

In the book, I examine current and emerging remedies and solutions to reverse Climate Change. Green, electric, emissions-free travel by car, plane and other fascinating new forms of green transportation are showcased in the book.

Green Renewable Energy

The book presents the latest renewable energy solutions being deployed and developed. New energy sources include hydrogen, wind and solar power. We look at the question of how global utility companies are supporting or not supporting the use and development of renewable energy sources like wind and solar power. According to an Oxford University comprehensive analysis of 3,000 power companies worldwide, an overwhelming majority are not supporting and funding renewables. In fact, the Oxford study concludes global utilities "are undermining the fight against Climate Change".

Green Governments, Businesses and Investing

The book also details the efforts of governments worldwide to regulate, reduce and eventually eradicate greenhouse gas emissions. The green goals of major corporations like Apple and Microsoft are showcased. As are the very specific green investment parameters newly instituted by global financial institutions, such as Blackrock.

If you care about Climate Change's relentless assault on the Earth, the air we breathe and the water we drink and the growing global fight to combat Climate Change, you will find this book of great value.

TABLE OF CONTENTS

AUTHOR'S BIOGRAPHY

Ed Kane created and serves as Executive Producer of CEO Global Foresight, a national news program on PBS focused on break-through innovations. He is the author of 24 books on the latest innovations across industries. Ed is a science graduate of the University of Pennsylvania.

1. Big Economic Benefits of Fighting Climate Change

Source: Climate Change Stock Image

NEW SCIENCE

A study published in the scientific journal Nature finds that if nations don't adhere to the 2015 Paris Agreement to keep global warming below 2 degrees Celsius and significantly cut CO2 emissions, the global economy will lose at least $150 Trillion to as much as $792 Trillion by the end of the century. Co-author of the study Biying Yu of the Beijing Institute of Technology says the general consensus in the global academic community is that climate change may lead to a global catastrophe with huge socio-economic loses. Unfortunately, most of the 190 nations that signed the Paris Accords are falling far short of their commitments.

Calculating A Green Future

Yu and her team found that if nations optimally cut their CO2 emissions, net global economic benefits would range from $127 trillion and $616 Trillion by 2100. The dollar figures are derived from climate change damage avoided minus the costs to cut global warming. The large range is due to simulating strategies, technology developments and climate damage.

Global Cooperation Needed

This study is so specific that it provides for each nation exact emissions reduction targets, economic breakeven points and top costs of various strategic approaches to fighting emissions and climate change. It clearly demonstrates the benefits of taking aggressive action against Climate Change both for the Earth and the global economy. The study concludes combating Climate Change is not a matter of one nation's actions. It takes collective action and cooperation from countries around the world.

2. Greenland Ice Sheet Record Melt

Source: NASA Greenland Ice Sheet Melt

New Study: Worsening Climate Change

The signs are undeniable that Climate Change is accelerating and breaking scientific estimates. A new study has documented that the Greenland ice sheet lost a record amount of mass in 2019. The loss is so dramatic that scientists are expected to start restating and redefining their worst case scenarios based on the new Greenland findings. The worst case scenarios are for rising ocean waters and flooding affecting millions of people in coastal locations.

New Melt Shatters Records

The rate of Greenland ice sheet melt, caught by satellites and analyzed by scientists through derived data, is record shattering. Geoscientist and glaciologist Ingo Sasgen of the Alfred Wegener Institute in Germany documented the loss of 532 gigatons of ice, or 66 tons of ice for every person on Earth - in 2019 just in the Greenland ice sheet. Greenland experienced much warmer temperatures in 2019. The 2019 melt is 15% more than the previous record set in 2012.

66 Tons of Ice Melt for Every Person on Earth in 2019

Clearly, Greenland's ice sheet is melting more quickly than expected. Experts say a total melt of the Greenland ice sheet would raise sea levels by at least 20 feet. Sasgen believes we are on a path of accelerating sea levels. If the 2019 rate of melt in Greenland persists at the level of 66 tons of ice for every person on Earth year after year, coastal flooding will affect up to 30 million more people each year by the end of this century.

3. Climate Change's Wrath

Source: Napa Valley, CA AP

From California to Alaska, Greenland and China

There is growing evidence that Climate Change and Global Warming are worsening. In California, hundreds of wildfires are burning out of control in the very late summer of 2020. They are being spawned by blistering heat and record numbers of lightning strikes. There were 11,000 lighting strikes in a few days, torching trees, brush and triggering the massive wildfires. 500,000 acres have been burned, communities are threatened, buildings have been destroyed and tens of thousands are evacuating. The Bay Area had the worst air quality in the world in late August 2020.

California Governor Gavin Newsom commented: "If you are in denial about Climate Change, come to California."

China's Torrential Rains

Meanwhile, in China torrential rains and massive flooding caused the evacuation of a million people and caused billions of dollars in damage. China's huge Three Gorges Dam experienced its highest water levels ever following months of torrential rain in the region.

Alaska's Shrinking Salmon

In Alaska, scientists have reported that wild salmon are significantly decreasing in size. For example, Chinook salmon are 8% smaller than they averaged in 1990. Scientists say the likely cause is warmer oceans from Climate Change. They add the wild salmon population is threatened along with coastal communities and the indigenous people who depend upon them for a livelihood.

Greenland Melting

The Greenland Ice Sheet is melting at unprecedented rates. New science documents 532 gigatons of ice was lost in Greenland in 2019. That's the equivalent of 66 tons of ice melt for every person on Earth. The pace of Climate Change is unrelenting and its path of destruction spans the globe.

4. UK Legally Binding Environmental Targets

Source: London Tower Bridge

Fighting Climate Change

Britain is rolling out legally binding environmental targets to combat climate change. The targets include cleaner water, reduced air pollution, waste reduction and more biodiversity. The legally mandated targets are designed to improve the environment and rebuild the economy. The new legislation, set to be considered by Parliament, will force the current and future British governments to focus on improving the environment.

Build Back Green

British Prime Minister Boris Johnson has pledged to "build back green" as the British economy was cut by 20% in the second quarter of 2020 by COVID-19 lockdowns and shutdowns. A new watchdog agency "The Office for Environmental Protection" is being established to monitor the UK's progress toward the environmental targets. The Office will also oversee the government's progress toward its commitment to have the UK be net-zero emissions by 2050. Britain was the first G7 nation to commit to that 2050 net zero-emissions target. And, it will be hosting the UN's Climate Change Summit in November 2021. The 2020 Summit has been postponed because of COVID-19.

5. Japan's Flying Car Takes Off

Source: SkyDrive test flight

Japanese Govt. Wants Flying Cars Operating by 2023

The beginning of the flying car era just got a lot closer. Japan's flying car, the SkyDrive, very recently took off with one person onboard in a successful test flight. It lifted up several feet (1 to 2 meters) and then hovered in a netted area for four minutes. It was a modest test flight but the vehicle worked.

Crowded Skies

There are more than 100 flying car projects around the world. The most notable are the Dutch PAL-V, Lithium of Germany, Joby Aircraft of California, Wisk (Boeing & Larry Page's Kitty Hawk) and SkyDrive. According to SkyDrive leader Tomohiro Fukuzawa, only a very few projects have succeeded to fly with a person onboard. The Japanese government wants flying cars in the skies over major Japanese cities to ease traffic by 2023. Fukuzawa expects SkyDrive to be an available product on the market by 2023.

eVTOL

SkyDrive is an eVTOL, meaning an electric vertical take-off and landing vehicle. That means you don't need an airport. It's point to point travel, for instance, from your driveway to the mall. eVTOLs promise green electric personal air mobility over traffic jams. Right now, the SkyDrive can fly for 10 minutes. The company says when it reaches 30 minutes worth of fly-time, it will have great potential for commercialization, including as exports to places like China.

Next Steps

To go into very successful commercialization, flying cars need to be reasonably priced. Advanced air traffic control systems need to be in place along with other needed infrastructure and the vehicles need to have good range on a single charge. In Japan and elsewhere these requirements are being worked on. The Japanese government is pushing hard with a plan for flying cars as a business service by 2023 and greatly expanded commercialization by 2030. SkyDrive has been well funded over the past few years with backers like Toyota and Panasonic among others.

6. Fish Finding Drone For Fishermen & Nature Lovers

Source: Chasing Innovation

Green Aquatic Drone with Camera - Controlled by SmartPhone

The Chasing F1 Fish Finder drone is a first. It's an aquatic drone, powered by lithium batteries, to explore the underwater environment. The user can operate it from shore or on a boat. The drone drops its tethered camera into the water to find, watch and track fish. The user operates and controls the drone system by their smartphone. Chasing F1 relays live pictures of its underwater fishing locations and can also provide videotapes and still images. It's the invention of the Chinese company Chasing Innovation, which specializes in underwater drones. It sounds like the answer to many a fisherman's prayers, It also provides new, live, taped and photo views of underwater life for nature enthusiasts.

Controlled by the User Via Their Smartphone

The fish finding drone is wirelessly controlled by WiFi up to a distance of 98 feet by an iOS/Android app on the user's smartphone. Four thrusters enable the drone to move forward, backwards, sideways and circle a location. There's a hook in the rear of the drone to tow a sonar unit or a remotely controlled bait dispenser. The top half of the drone remains above water. It's painted yellow and clearly visible for the user to pinpoint where the drone is and the best fishing location. It also has GPS which shows its location on the smartphone GPS app. If you just love to see fish in their natural habitat, it's a great new piece of environmental technology.

Tech Specifications

The drone/camera system can go down to a depth of 92 feet. On the app screen, the fisherman or nature enthusiast can see exactly what the camera sees along with the depth and water temperature. The camera can beam infrared light to make murky water locations visible. Every function of the Chasing F1 is wirelessly controlled by the user. It's a green water machine, powered by a 4,800 mAh lithium battery. It works 4 to 6 hours on a charge. The cost is $699.00 and it's available in the US and elsewhere.

7. Mysterious Bullet Plane: World's Most

Fuel Efficient Aircraft

Source: Otto Aviation Celera 500L

Celera 500L

A mystery, bullet shaped aircraft, long the subject of speculation, has finally been revealed. It's called the Celera 500L and it's the invention of California based Otto Aviation. Until now, the development of the plane has been totally under wraps. It's being called the world's most fuel efficient plane.

Incredible Fuel Efficiency

The Celera 500L is a six person private jet that flies at the speed of a jet but with 8 times lower fuel consumption. Add to that the range is double that of a similarly sized aircraft. The company says 31 successful test flights have been performed. They claim the Celera 500L is the world's "most fuel efficient, commercially viable aircraft in existence."

Specifics

This jet runs at 18 to 25 miles per gallon of fuel economy versus 2 to 3 miles per gallon of fuel economy for a regular jet. It can reach speeds of 460 mph, has a range of 4,500 nautical miles and the $328 hourly operating costs are six times lower than a comparable jet. The company says it can service virtually any city to city route in the US without refueling.

How Can It Provide This Kind of Performance?

The magic is the smooth bullet shape and it comes down to laminar flow. That is the minimum drag amount for the aircraft's surface and allows for smooth layers of air flow. The plane's aerodynamic airframe requires a lot less horsepower to operate. Its RED A03 engine can run on biodiesel or Jet A1 fuel. First commercial deliveries are expected in 2025. The aircraft now needs to go through the certification process.

8. Oxford University Global Study Says Majority Are Not

Source: NASA

Power Companies and the Fight Against Climate Change

A new study by scientists at Oxford University in the UK suggests that power companies are very slow to support and utilize green, renewable energy sources like wind and solar power. In fact, of the 3,000 power utility companies around the world that the Oxford researchers studied, only 10% of the energy suppliers have prioritized renewables over fossil fuels. Many continue to invest in fossil fuels. The study concludes that utilities' slowness to embrace renewable energy sources is undermining the fight against Climate Change.

Europe Coming on Green

Europe is ahead of the curve in using renewables. For instance, 40% of the United Kingdom's electricity is coming from wind and solar. But globally, many of the new wind and solar instillations are being built by independent producers not by utilities.

Comprehensive 20 Year Study and Analysis

The Oxford University study used machine learning techniques to analyze 3,000 utilities' activities over the past 20 years. According to the study, most utilities are sitting on the fence about renewables. Globally, only 1 in ten have expanded their renewable based power generation more quickly than their oil and coal generated power, which is a major contributor to greenhouse gas emissions globally. The Oxford study raises troubling concerns about the commitment to renewable energy sources by the majority of power companies around the world.

9. AI to Predict Climate Change From Microsoft & Partners

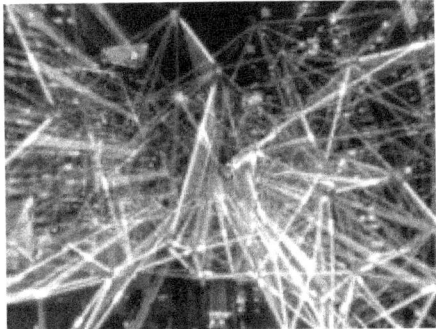

Source: DARPA AI

Targets Include Extreme Weather, Natural Disaster Predictions

Microsoft and the US Department of Energy (DOE) are spearheading a new partnership to create artificial intelligence (AI) tools capable of predicting natural disasters like the extreme weather events and historic California wildfires experienced from Climate Change. The AI partnership includes the Pentagon's Joint AI Center, the Pacific Northwest National Lab along with Microsoft and the US Department of Energy.

Climate Change and Extreme Weather

One of the most exciting applications of AI is to forecast and predict Climate Change and extreme weather conditions, like the historic flooding in China, the record breaking back to back hurricanes hitting Louisiana and the blistering heat much of the world endured in the summer of 2020. Global weather services have satellites, computer modelling and tons of data on weather conditions, patterns and developing storms. Plugging all of that data into deep learning machine models (AI) will likely result in earlier and more accurate predictions of extreme weather, identify important climate change patterns and significant developments.

First Five Consortium

The DOE calls the new AI initiative the First Five Consortium. The name refers to the critical first, five minutes after a disaster strikes when first responders make critical decisions. The AI tools developed will help provide breakthrough information for that decision making. The DOE-Microsoft partnership will develop and deploy AI in four critical areas:

- Wildfire prediction and fire line containment - of great and growing importance in California, the Amazon and elsewhere
- Natural disasters like predicting Earthquakes and Volcanos
- Search and rescue such as spotting survivors in burning, collapsed buildings
- Damage assessment - knowing Arctic ice melt rates much sooner and predicting the consequences more quickly would be a critically important application.

This is a great private-public partnership putting the breakthrough tools of AI at work for humanity.

10.   World's First Solar Sky Dive

Source: Solar Stratos

Flying on Sunshine, Parachuting Through the Air

It's the world's first solar skydive. Parachutist/pilot/adventurer Raphael Domjan safely completed the world's first jump from a solar powered, electric plane over Switzerland. The plane soared to an altitude of 5,000 feet. In his descent, Domjan reaches speeds of 93 mph (150 kilometers per hour). The flight and dive were designed to showcase the power and potential of solar, electric powered aircraft.

SolarStratos Project

Domjan is the founder of the SolarStratos project, which organized the historic Solar SkyDive. He co-piloted the plane during its ascent. He is the first parachutist to jump from an electric plane. He believes that flying in green, electric planes will change forever the future of the sport for skydivers.

Next Takeoffs

In 2022, the SolarStratos team hopes to make more history. They aim to fly a solar powered, electric plane to the stratosphere and reach an altitude of 65,617 feet (20,000 meters).

11.　　Solar, Autonomous Hydrofoil

Yacht

Source: Nemesis Yachts

Autonomous Wing Sailing

Nemesis Yachts has created the ultimate solar powered, autonomous super yacht - the Nemesis One - with zero emissions. The yacht is powered by the sun, propelled by electric and hydrogen energy and is totally autonomous. The company says it is the world's first high speed, autonomous, hydrofoil luxury catamaran. It operates silently with no fumes and no emissions. This concept is a breakthrough piece of ocean sailing technology, incorporating solar, hydrogen and electric energy breakthroughs.

Zoom in on Specifics

The yacht is huge: 332 feet in length and the hi-tech, automated sail is 292 feet high. It has a cruising speed of 35 knots and top speed of 50 knots. It zooms through the ocean by autonomous hydrofoils and the autonomous wing sail. There are different interior modular configurations that can be removed for instance to go into an ultra-light racing mode.

Automatically Adjusting Wing Sails to Conditions

Nemesis Yachts specializes in multi-hull ships that are fast, comfortable and emissions free. Nemesis One has a carbon fiber build. If it goes from concept to reality by someone ordering it, the catamaran will be the world's fastest luxury hydrofoil yacht. According to the company, the wing sail automatically adjusts to conditions. There is no need for manual handling. The company says the wing sail is 2&1/2 times more productive than a traditional sail.

Data-Driven Lidar Equipped Vessel

Both the sail and the hydrofoils adjust through data collected and analyzed by next generation Lidar sensors that capture the yacht's surroundings. The data includes real time ocean conditions, like wave heights, shape and frequency. Concerning price, there are estimates of $90 million to have a Nemesis One for your own adventure into very high tech sailing and racing.

12. China, Climate Change and Investing

Source: Chinese Stock Exchange

Promoting the Environment, Social and Governance Standards

MSCI, the New York City based financial services company, has launched Climate Change indexes for China stocks. The indexes allow investors in China to put their money into companies with lower carbon emissions. The indexes allocate more weighting to less polluting companies, such as the internet giant Tencent and LONGi Green Energy Technology. MSCI has launched similar Climate Change indexes in the EU, US and Japan. MSCI is a global provider of equity, fixed income, hedge fund stock market indexes and analysis tools.

Megatrend: Investors Focusing on Climate Change

MSCI expects global investors to increasingly focus on Climate Change and put their money into companies with strong pro-environment policies and results. MSCI is in discussions with a number of institutions about creating investment products based on Climate Change indexes. In China, the launch of the Climate Change indexes comes at a time when the Chinese government is trying to open up its capital markets. China is involved in a very difficult balancing act between cutting pollution and moving its economy forward. Like the rest of the world, the Chinese economy has been hard hit by COVID-19.

ESG Standards

MSCI says it's trying to promote environmental, social and governance (ESG) standards in China. The new Climate Change indexes for China are based on the MSCI China Index and MSCI China A Index, which have been operational for several years here in the US. The company says the market will discount share prices of companies with poor ESG compliance. Currently, ESG-focused foreign investors are putting pressure on listed companies in China, where ESG compliance has been seen as inconsistent and sub-standard.

13. Germany's First Green Bonds for Environment

Source: Germany's Hydrogen Powered Green, Clean Train

Funding Clean Transportation & Renewable Energy Projects

For the first time, the government of Germany plans to issue $13 billion in so-called Green Bonds to fund renewable energy and clean transportation projects. The $13 billion worth of bonds will be issued this year - 2020. Germany is tapping into the financial markets to increase its commitment to fight climate change. This is a case of putting money up against specific environmental targets to fight climate change. Germany has a strong green commitment but it is even getting greener.

September Launch

The first issuance happened in September 2020 of a 10 year bond or Bunds that is worth 4 billion euros in value. The rest will be issued in the fourth quarter of 2020. By 2023, Germany wants to spend 54 billion euros as part of its climate change agenda. The agenda includes introducing a carbon tax to cut greenhouse gas emissions by 55% by 2030 as compared to 1990 levels. By the Green Bond issuance, the German government says it want to show the public how green and climate friendly initiatives and targets can be made "transparent and predictable".

Global Green Bond Market

Using green bonds to fund environmental projects and fight climate change is growing. It is most popular in Europe. Germany follows Poland which launched the world's first green bonds in 2016. France issued its first green bonds in 2017 and it now is the world's leading issuer. Germany is a leader in supporting clean transportation, like hydrogen trains, eVTOL flying cars, and renewable energy. And, now Germany is issuing Green Bonds to accelerate, underwrite, guarantee and increase its environmental projects. Overall in 2019, green bonds accounted for close to 3% of global bond issuances or $205 billion.

14. Microsoft Takes on Climate Change

Source: Microsoft

Commitment from the Top

Microsoft has joined the fight against Climate Change in a big way. It announced that by 2030 it will have removed more carbon from the atmosphere than it emits. And, by 2050, it hopes to have taken out as much CO2 as all the direct emissions the company has made since its founding in 1975.

Very Different, Innovative Approach

The world's largest software company's approach to fighting Climate Change is quite different and innovative. Most companies promise to cut ongoing emissions or prevent new ones. By contrast, Microsoft's pledge to remove existing carbon from the atmosphere is unique for corporate America.

CEO Leadership

The commitment to fighting Climate Change is coming from Microsoft CEO Satya Nadella. Speaking to employees at Microsoft's Redmond, Washington headquarters, he said we, as a global company, have to offset the damaging effects of Climate Change. He added if global temperatures keep rising, "the results will be devastating".

Specific Components of the Plan

There are several key components to the impressive approach that Microsoft is taking:

- By 2030, it wants carbon emissions cut in half across its entire supply chain.
- It's creating a "Climate Innovation Fund" to invest $1 billion over four years to accelerate development of carbon removal technologies.
- It's greatly expanding an internal fee it's been charging its internal business groups since 2012 for their carbon emissions including use of electricity, air travel and all other work related carbon emissions.

Microsoft President Brad Smith says that to accomplish their goals by 2030, technologies that don't fully exist have to be developed. Microsoft expects to release 16 million metric tons of CO_2 in 2020 including for corporate travel. Their aggressive carbon removal plan and transparency about their own carbon emissions are impressive in the fight against Climate Change.

15.   US Renewable Energy Investments Record-Breaking

Source: Wind, Solar Energy Stock Image

Wind and Solar Lead the Way

Despite the Trump Administration's views on Climate Change and renewable energy, US companies broke records in their investments in green technology in 2019. Investments hit $55.5 billion, which is up 28%. That catapulted the US to second place in renewable energy investments behind China, whose financial bet on renewables fell 8% to $83 billion. Clearly, China remains the far ahead leader in pushing renewables. Europe's renewable energy investments fell 7% to $54 billion, right behind the US. Interestingly, Brazil's financial investment soared by 74% to $6.5 billion. The data was collected by Bloomberg.

New Green Energy Technologies

The big investment push in the US for new renewable energy technologies and expansions is by wind and solar energy companies. They're rushing to qualify for federal tax credit that ends by 2021. There is growing sentiment across the US that renewables are preferable, clean and green energy sources. Many electric utilities are investing in new renewable energy technologies. Rhode Island has announced plans to have 100% renewable energy power by 2030. And tech giant Apple wants to be powered by renewables globally. Renewable energies are getting a big green light.

16.   Adidas Ocean Plastics Recyclable Fabrics

Source: Adidas Ocean Waste Commitment

Two New Fabrics 100% Recycled

German based sports gear giant Adidas is going very green and blue. It has introduced two new fabrics: Primeblue and Primegreen. Both are composed of 100% recycled polyester. Primeblue contains Parley Ocean Plastic, which is a fabric made from recycled ocean plastic waste. Primegreen, contains only recycled plastic and recycled polyester. Adidas has showcased some running shoes made of recycled ocean plastics. It is greatly widening its environmental commitment.

Very Specific Environmental Goals

For 2020, Adidas is committed to having 50% of the polyester it uses in products recycled polyester. By 2024, it will use only re-cycled polyester. And, it commits to cutting its carbon footprint by 30% by 2030 and to be carbon neutral by 2050.

Green Running

It is also making, for instance, Futurecraft Loop running shoes that are fully recyclable. Adidas' goal is to make products that can be reused and remade before returning to nature. That certainly makes for cool, green and blue running.

17. 2020 World Economic Summit Focused on Climate Change

Source: World Economic Forum

Several Global Corporations Rising to the Challenge
The 2020 World Economic Forum in Davos, Switzerland focused on sustainability, sustainable business and accelerating action globally to thwart Climate Change. There were more than 2,000 participants including 113 billionaires. Interestingly on the eve of the Forum, several global companies announced detailed plans to combat Climate Change.

Nestle, Amazon, Microsoft, Black Rock
Swiss food giant Nestle is investing $2 billion to use recycled

plastics and cut its use of new plastics by 1/3rd by 2025. It's an impressive move by the global food leader to reduce global plastic pollution. Amazon announced that it's going carbon neutral by 2040. As part of that, it has ordered 40 electric vans from Deutsche Post's Streetscooter unit for deliveries in Munich. Meanwhile, Microsoft has committed to removing by 2050 the carbon emissions equivalent of its entire carbon footprint since it went into business in 1975. And, by 2030, it will have removed from the atmosphere more CO_2 than it emits. Finally, Black Rock, the world's largest asset manager, announced that it is making the fight against climate change a key part of the firm's strategy. Essentially, the firm with $7 trillion under management, is escalating its holdings in green, sustainable investment vehicles.

18.   Big Trend Sustainable Investing: $7 Trillion Point of View

Source: BlackRock's Sustainable Home Banner

BlackRock Goes Green

BlackRock, based in New York City, is the world's largest asset manager with $7 Trillion under management. CEO Larry Fink is making fighting Climate Change a key part of his investment strategy. His focus is sustainable investing. He says that the dangerous impact of Climate Change is causing investors "to reassess core assumptions of modern finance."

Concrete Steps

Fink has a strategic plan that he's implementing. Essentially, he's greatly escalating holdings in green, sustainable investment vehicles. Specifically, he's:

- Doubling BlackRock's sustainable ETFs to 150
- Putting pressure on index providers to include more sustainable benchmarks
- Selling $500 million worth of investments in thermal coal producers
- Voting against boards and management who don't account for climate risk or don't have a sustainability plan.

Investing Advice

As a global investment leader, Fink's words and actions are unprecedented. For investors this dramatically underscores important new opportunities in green, sustainable investing.

19. France Bans Fossil Fuel Cars by 2040

Source: Traffic in Paris, Stock Image

More Mobility, Less Carbon Dioxide
The goal of the government of France is to be carbon neutral by 2050. The government of President Emmanuel Macron plans a ban on gas and diesel (fossil fuel) powered cars by 2040. The government will ban their sale by 2040 in order for France to become carbon neutral by 2050. This bold initiative is part of the government's new law on mobility to help the environment.

Green, Alternative Fuels
The ban on fossil fuel cars demonstrates President Macron's commitment to the environment. The French government also plans to help its car makers go electric, hydrogen and biofuel. The deployment of electric vehicle charging stations will also be facilitated. That includes an innovative provision to allow the right to apartment building dwellers to ask for EV plugs in their parking lots.

Fewer Traffic Jams, Innovative Approach to More Mobility
Other innovative components of the French Mobility Law include offering employees a tax-free, 400 euro subsidy to go to work by bike or car pool. France is also providing legal structures for free-floating scooters, bikes and car sharing to increase mobility without contributing to climate change.

20. US CO2 Emissions Drop Dramatically

Source: NYC April 2020

US Energy Information Administration Data

COVID-19 lockdowns, travel restrictions and working from home in the US have resulted in significantly reduced CO_2 emissions. Specifically, in April 2020, CO_2 emissions in the US hit a record low - the lowest level since the US Energy Information Administration (EIA) started keeping CO_2 emission records in 1973. The dramatic reduction is all about significantly decreased travel by Americans to avoid exposure at work, school, on public transit and planes to the COVID-19 virus. If there is any positive out of this horrific pandemic, the reduction in CO_2 pollution is it.

Gasoline and Jet Fuel Use Plummeted

The ripple effect is clear. Stay at home advisories and orders cut travel and petroleum use. It also shifted electricity consumption from commercial and industrial users to residential users, according to the EIA. The EIA documented that during April 2020, gas consumption was way down. Gas is the most used petroleum fuel in the US. Jet fuel use was also significantly down.

Zoom In on the Numbers

CO_2 emissions in the US fell to 307 million metric tons in April, that is down 25%. Airline travel was down a whopping 96% from April 2019. And US road travel was down 40%. The EIA forecasts that CO_2 emissions will start rising again in 2020 but by the end of the year will still be down by 11% when compared to 2019.

21. India's Solar Ferries for Commuting

Source: Kerala State Water Transport Department

Future of Travel

India's solar ferries are a big part of the future of green, clean water travel. Their solar ferry, the Aditya, carries 1700 passengers daily on a two mile route from a lakeside community back and forth to the town of Vaikom. India's first solar powered ferry has merited an award from the international electric boat Journal Plugboats. It generates zero pollution and is 30 times cheaper to run than the diesel ferry it replaced.

Solar Power Delivers

This new solar water ferry delivers. It runs on 70 kilowatts of electricity, 65 % of which come from the boat's solar panels and the rest from the grid. India's Kerala State Water Transport Department has commissioned two more solar ferries and a solar cruise ship. It's also converting its 48 diesel ferries to solar powered ferries within a few years.

Solar Powered Water Travel

The Aditya is considered a solar ferry prototype. When compared to diesel ferries, it runs at one half of the cost of operation with the same performance. It could have applications to solar boats operating in Europe, Africa, Asia and elsewhere. The operators say it is very easily replicated. It offers a green clean future of solar powered water travel for thousands, if not millions worldwide, by ferry.

22.   Japan Wants Flying Cars by 2023

Source: SkyDrive

Japan's Traffic Mitigation Goals

The government of Japan is determined to deploy flying cars for urban traffic relief. It wants flying cars soaring over its big cities by 2023. It believes flying cars will be a solution for traffic congestion and transportation problems in major cities like Tokyo. Japanese government leaders also view them as a solution for problematic mountainous and remote regions where access is difficult.

Japan's SkyDrive

There are several notable flying cars/flying taxi eVTOLS (electric vertical takeoff and landing) vehicles under development. Players include Airbus, Uber and Boeing with their prototype eVTOLs. And Japan is developing several of its own flying cars. Of interest is the newest Japanese flying car startup SkyDrive with its SD-XX, 2-seat flying car. This eVTOL can be parked in an office parking lot in just two spots.

SkyDrive Takeoff

Tokyo-based SkyDrive has raised 1.8 billion yen ($17 million) from investors. It's working with Toyota City and using its indoor development base and testing facilities. The vehicle is a piloted vehicle with 2 propellers set in 4 corners of the vehicle. It's powered by an electric battery. Engineers hope to launch a commercial version with speeds up to 37 mph and with a range of 19 miles.

SkyDrive Flying Cars Coming Soon

The Japanese government wants to launch air taxi services in 2023 and have flying car companies like SkyDrive start selling autonomous flying cars by 2028 to the general public. Flying taxi service companies are targeted to start first in Tokyo and Osaka in 2023.

23. China's Revolutionary Green Travel Plans

Source: China's Maglev train under development

Smart Ships, Smart Trains, Smart Autonomous Vehicles

According to China's Ministry of Transport (MOT), the country is developing very high-speed Maglev trains capable of travelling 600 kmph (373 mph) and high speed passenger trains capable of travelling 400 kmph (249 mph). Models of the Maglev were launched in summer 2020 at a factory in eastern China's Shandong Province. China has revolutionary plans for its transportation system. It is developing a world-leading transportation network loaded with smart and digital technologies, including 5G, artificial intelligence and remote sensing satellites.

2025 Transportation & Energy Sources

The Chinese government's blueprint is substantial. By 2025, it intends to expand the use of smart ships, smart boats, smart trains and smart autonomous vehicles. The plan also includes new construction of green energy infrastructure to cut emissions, pollution and save energy. For instance, on highways, China's MOT says they will deploy solar power generation facilities and ultra fast electric vehicle recharging stations. The MOT will also encourage the use of "clean bunker fuels" like liquefied natural gas. And in ports, they will develop wave/shore power technologies.

On Track For a Green Future

The Chinese government expects to fund these smart transportation and green energy projects with government funding and private capital. This is China's vision and blueprint to lead the world in transportation and green energy innovation in its transport network by 2025.

24. Trump Ok's Drilling in Arctic Natl. Refuge

Source Arctic National Wildlife Refuge Polar Bears & Caribou

Wildlife Refuge at a Time of Climate Change

President Trump has finalized plans to allow oil and gas drilling in the Alaska Arctic National Wildlife Refuge (ANWR). No drilling has occurred there in decades. The refuge is home to polar bears, porcupine caribou and other species deeply threatened by climate change, which is melting ice in the Arctic at a record pace. President Trump plans to issue long term leases to oil and gas drilling companies before there is any potential change in US Presidential leadership, come January 2021.

Politics and the Environment

Alaska Governor Michael Dunleavy and the energy industry say this move will create jobs and boost the economy. Democrats and environmentalists say it will do destruction to the Arctic's unique ecosystem and the native Alaskans who are dependent on the region. Add to that, the massive disruption to the lives of the magnificent animal species that call it home. Those who disagree with President Trump's move to allow drilling in one of Alaska's most pristine regions say this is a big political giveaway to Big Oil on the eve of the November 3, 2020 National Presidential election.

Interior Department Moving Forward

Interior Secretary David Bernhardt says he can hold a sale of oil and gas leases in the ANWR by the end of this year. If he does, oil and gas production could start in 8 years and last for another 50 years. Environmental groups threaten to sue. Bernhardt says the Interior Department procedures have been so air-tight, he doesn't think this can be challenged legally. This new environmental battlefront is seething and it will likely face major legal challenges from many stakeholders.

25. Cadillac is Leading GM's EV Lineup

Source: Cadillac Lyriq

Taking on Tesla

Cadillac has unveiled its first fully electric vehicle, the Lyriq. The EV is designed to challenge Tesla's leadership in electric cars. The Lyriq is powered by Cadillac's new Ulteum battery technology and provides a range of more than 300 miles on a charge. For GM, Cadillac is its leading electric vehicle brand. Led by Lyriq, Cadillac says it will roll out a new portfolio of transformative EVs during the 2020's decade.

Big Electric Rollout

Cadillac says it will roll out 22 electrified vehicles by 2023. Lyriq, with its sleek, crossover design, will first launch in China and then in the US. Production is scheduled to start in 2022. The company says pricing will start at $75,000 to $100,000.

Signature Features

The Lyriq has several signature features. Its Super Cruise semi-self-driving technology provides full, handsfree operation on over 200,000 miles of GPS mapped highways. It also has a huge, 33" LED screen that stretches across the dashboard and can display more than 1 billion colors. And, Cadillac says the vehicle has lighting choreography with a "black crystal front grill" and slim LED lighting. A split tail light design is also a distinguishing feature of the Lyriq, GM and Cadillac's first big salvo into the EV model wars.

26.   Secret Lives of Sharks

Source: Reef Shark Image

Sharks Have Daily Gathering Times to Share Information

New scientific research indicates that sharks have secret lives. They form social network communities. Some species of sharks have social lives that endure for years. The solitary ocean predators may not be so solitary. A team of researchers, led by Florida International University marine biologist Yannis Papastamatiou, tracked with technology, the daily habits and social behavior of 41 reef sharks in an Atoll 1,000 miles southwest of Hawaii. What they found is amazing. Using acoustic transmitters to track the animals and camera tags to see the interactions, they documented that the sharks form social networking communities on a daily basis that can last for years.

Social Shark Communities

Some of these shark communities remained together for the four years of the FIU study. The sharks spent mornings together in groups of 20 in the same part of the reef, dispersed for the rest of the day and night and then re-congregated morning after morning. Reef sharks are medium sized at 6 feet in length. The researchers say other sharks species may be solitary. But, the reef shark community networks are not about nesting, vocalizing or friendly interactions, like such community gatherings are with birds and some animal species. The scientists think these community daily gatherings may be informational and provide individual shark members with the location of food prey for the day. These predatory animals are a lot more complex and smarter than we ever imagined before.

27.   Zero Emissions Airlines by 2030

Source: Airbus Hydrogen Plane Concept

European Union's Hydrogen Strategy

In Europe, there is growing consensus building around utilizing hydrogen as a primary, green, renewable energy source. The momentum is coming from the both the European Union and global aerospace giant Airbus. France-based Airbus believes that hydrogen fuel is the quickest means to turn commercial passenger jets into zero emissions passenger flights. Company leadership forecasts that hydrogen airliners could be flying in much greener skies by the early 2030's.

Hydrogen Agenda

To develop and deploy hydrogen commercial jets by early 2030, Airbus Vice President for Zero Emissions Technology Glenn Llewellyn recommends utilizing and reconfiguring automotive and space launch hydrogen technologies that are already developed to create hydrogen commercial passenger jets. Toyota is a major developer of hydrogen fueled cars and the Japanese automaker also advocates for a hydrogen powered society. Hydrogen developed from renewable energy sources is clean, green and has zero emissions. In the space industry, the Vulcan 2 liquid fuel engine rocket is hydrogen powered. The rocket lifts the Airbus designed Ariane launcher into space for the European Space Agency. Llewellyn's strategy is to jumpstart the development of hydrogen aircraft by utilizing proven, existing hydrogen technologies.

European Union

The European Union has developed a hydrogen strategy to move forward the production, use and distribution of hydrogen throughout the region. There are even proposals to create a hydrogen pipeline throughout the European Union by tying together existing gas pipelines and moving green hydrogen through it to major population and industrial centers for use as green fuel for everything from commercial hydrogen jets, hydrogen cars, hydrogen trains and more.

28.   Global Race to Drive on Hydrogen Fuel

Source: Hyundai XCIENT Trucks for Switzerland

US Startup Nikola Coming on Strong

South Korea based Hyundai is quickly becoming the leader among global automakers in deploying hydrogen powered vehicles. It recently shipped the first ten of its Hyundai XCIENT Fuel Cell hydrogen semi-trucks. It's the world's first, fuel cell powered, heavy duty truck. And it's the first, mass produced hydrogen truck. The hydrogen vehicles went to Switzerland. Another 40 hydrogen trucks from Hyundai will be on Swiss roads by the end of 2020. It's the beginning of a Hydrogen Highway Ecosystem, with hydrogen fuel stops eventually being built into the highway infrastructure. It's the future of zero highway emissions.

Hydrogen Highway Ecosystem

The Hyundai hydrogen semi's have a 255 horsepower fuel cell converting hydrogen into electricity. They have a driving range of 250 miles on a tank of hydrogen. To make the hydrogen transportation work, a highway ecosystem of hydrogen fuel stops needs to be built into the infrastructure. In Switzerland, we're seeing the beginning of hydrogen powered big trucks. The emissions are nothing but water. It's a green, clean, zero-emissions way to drive. Experts believe the most powerful, first wave of hydrogen powered vehicles are big trucks. Cars, trains and planes are next.

US Startup Nikola

US hot startup Nikola's proprietary hydrogen/electric trucks have huge orders, to the tune of $10 billion. The company would like to collaborate with Hyundai on hydrogen technologies to accelerate them to global markets. Nicola is putting its electric trucks into production in 2021 and its hydrogen fuel cell electric trucks in 2023.

What Are the Prospects?

Hydrogen fuel is the holy grail. It is clean, green, efficient, zero emissions and eliminates the hours needed to recharge a huge battery for an electric truck or time to recharge an electric car. Hyundai would like to sell its hydrogen systems to other automakers. But Nikola is becoming a direct competitor to Hyundai's commercial truck business. So, time will tell about such a global collaboration. In the meantime, Hyundai has partnered with Audi on hydrogen vehicle technology. And Hyundai and Toyota are the leading global automakers pushing the development and deployment of hydrogen technologies. Hydrogen is a big part of the future of travel.

29.   Electric Car that Goes the Distance

Source: Lucid Air

New Vehicle has Long Range

Electric vehicle startup Lucid, based in California, has a new vehicle that can go the distance. The Lucid Air sedan has an expected EPA rated range of more than 517 miles on a single charge. That makes it the longest range electric vehicle on the market today. In fact, it gets 115 miles more range than Tesla's current highest range vehicle, the Model S, which has an EPA rated range of 402 miles. Tesla was the first EV company to break the 400 mile range with a widely available electric car.

Proprietary High Tech Motors

Lucid CEO Peter Rawlinson credits the company's high tech motors, invented at Lucid, with the industry's leading EV range. And, if you're wondering when you can buy one, the final version of Lucid Air will be presented in late 2020. The Air will go into full production at Lucid's Case Grande, Arizona facility. First cars will be delivered to customers early in 2021, according to the company.

Tesla Competitors Lining Up

Lucid claims the Air can go from 0 to 60 mph in 2.5 seconds. The company is expected to detail battery pack size and pricing once the final version of the Air is debutes. Lucid joins a number of startups like Rivian and Nikola in competing with Tesla.

30.   Innovation Turns Seawater to Drinking Water

Source: Quench Sea

Addresses Global Lack of Drinking Water

A new invention called Quench Sea converts seawater into clean, fresh, drinking water. The light, portable desalination device is the invention of London-based startup Hydro Wind Energy. The device removes salt, pathogens, parasites, contaminants and microplastics from the seawater. The company pledges to donate a device for every one they sell. They hope to have donated 100 million devices by 2027 which, they say, would impact a billion people globally in need of clean drinking water. The device is low cost - $60.00. It can produce three liters of water within an hour and it is manually powered.

Tremendous Global Need

2/3rds of the world's population live in water scarce areas and 7 million people a year die from water related diseases. The device has been independently tested and has been proven to work. It combines a hydraulic system, triple pre-filtration and a small, reverse osmosis membrane to desalinate seawater into fresh, clean water. The company says the device is for sailors, campers, travelers and for emergency use and humanitarian relief. Delivery of the innovative device is set for sometime in 2021.

31.   Tesla as a Global Green Energy Provider

Source: Tesla Solar Home

First New Target Market is Europe

Tesla is already the world's #1 electric vehicle automaker. Founder & CEO ELon Musk is offering a home energy package to consumers that would bundle solar panels and roofs, Powerwall energy storage and electric vehicle chargers. It's the whole package for going solar at home and driving clean, green electric vehicles. Tesla is currently surveying customers in Germany for their appetite of going very green and for their specific green energy needs. The bottom-line: entrepreneur Musk is attempting to add a brilliant layer of business onto Tesla electric vehicle operations as a global green energy utility to service his customers and new green customers' needs. He has petitioned to be granted licensing as a utility in the UK and already has a small grid system in Australia.

Specifics

Tesla is very methodical in its approach and in the survey wants to know how important to potential customers of its Energy Service Package the following components are:

- Supply to your home of clean, green energy
- Importance of home energy storage system
- Solar panel and solar roof needs
- Need for Tesla Wall Connector EV charging device
- Need for access to a public electric vehicle charging networks

Big Solar/Electric Expansion Plans

In the survey and marketing material Tesla is also mentioning the potential of grid services. CEO Musk wants to expand his solar and electric products globally. Currently, the solar panels and roofs are available in the US. This information gives you a view into how methodically Musk works as a business entrepreneur to expand his innovations into global markets. His approach is unique, bold and innovative.

32.   Europe's Hydrogen Energy Plans

Source: European Hydrogen Backbone Initiative 2020

Green Hydrogen Pipeline

A hydrogen pipeline stretching across the European Union, from Sweden, down to the south of Spain and to Sicily could radically transform Europe's energy system. That is the vision of European gas infrastructure companies of a "European Hydrogen Backbone" powering Europe. Their plan details transporting the new fuel across Europe via existing gas pipelines with minor modifications. They say this will provide a reliable means of transport.

Existing Pipeline Footprint

The existing pipeline network system would connect industrial centers that are capable of generating and utilizing hydrogen. It would also connect solar power plants and offshore wind farms into the pipeline network. The 11 transmission system operating companies say the hydrogen pipeline network could be totally completed by 2040 with major population centers brought on-line by 2035. They estimate the cost at $28.3 billion to $79.2 billion Euros.

European Union

The European Commission, which is the Executive Branch of the European Union, is bullish on hydrogen as a green fuel. They want green hydrogen (the byproduct of renewable energy generation with zero to low carbon emissions) scaled up to 1 million tons of production by 2024 and 10 million tons by 2030. They are also setting up funding sources to move their hydrogen plans forward.

33.   Hyundai's Big New EV Commitment

Source: Hyundai Kona

Adding Big EV Production Lines

Hyundai, the big automotive champion of hydrogen powered vehicles, has decided to also aggressively go electric. It's rapidly expanding its electric capacity to support selling $1 million electric vehicles a year by 2025 and also grabbing a global EV market share over 10%.

Hyundai's EV Offensive

Hyundai has plans to introduce two production lines dedicated to electric vehicles. One will go on line next year and another in 2024, according to an internal union newsletter obtained by Reuters. Hyundai is also engaged in discussions with Samsung to secure batteries and other electronic components for the EV's. EV car batteries are in short supply. Samsung is a major battery supplier to Tesla and GM.

EV Races Heating Up

Hyundai, which owns Kia, is the world's 5th largest automaker. It sold 86,434 battery EV's last year. That's more than VW at 73,278 but far behind Tesla at 357,500. Tesla is the world's leading EV automaker. It's off to the EV races.

34.   BMW's Big New EV Commitment

Source: BMW 5-series 2021

Big Green Lineup of Vehicles

BMW announced that it will offer a fully electric 5-series. This is part of a major product overhaul to offer green, emissions free vehicles. BMW wants to lower emissions both during the production of its vehicles and also when they are on global highways. It is even tying executive compensation to its climate goals.

BMW's 10 Year Plan

By 2030, BMW targets having 7 million plus electrified BMW brand electrified vehicles on the road, 2/3rds of which are fully electric. Their goals coincide with new European regulations that require carbon dioxide vehicle emissions be cut by 37.5% by 2030 as compared to 2021 vehicles.

BMW's E-Rollout

In 2021, BMW will have five, fully electric vehicles for sale, including the BMWi3, Mini Cooper SE, BMW iX3, BMW NEXT and BMWi4. By 2023, BMW will offer 25 electrified vehicles, including half that will be fully electric. BMW CEO Oliver Zipse did not offer a launch date on the fully electric series-5.

Tying Salaries to Climate Targets

BMW is going tough on meeting its electric vehicle targets. The company is holding management compensation to compliance with climate targets. And, it is committing to lower carbon emissions from its production sites by 80% per vehicle. That's a strong green environmental statement from BMW, one of the world's leading luxury automotive brands, headquartered in Germany.

35. Apple Going Totally Carbon Free

Source: Apple

Every Apple Product Will Be Carbon Free by 2030

Apple has made a global commitment to be carbon neutral by 2030 for all corporate operations, products, product manufacturing and its entire supply chain. The commitment is globally important as a corporate standard for a greener environment. Apple says every product, how it is manufactured and the energy used to produce it will be totally carbon neutral. Supply chain operations for Apple products will be totally carbon neutral. Apple commits to have a carbon free footprint globally by 2030.

Key Components

Apple has a comprehensive plan to deliver on its carbon-free commitment:

- **Products:** Apple will build its electronic products with low carbon and recyclable materials
- **Dave the Robot:** Dave is a big piece of the equation. Dave will disassemble, remove and recover plastic components from discarded iPhones, Macs, iPads and other Apple products to recycle them.
- **Supply Chain Clean Energy**: Apple suppliers have to up-

grade to renewable energy.

- **Apple**: Apple's corporate operations already run on 100% renewable energy on a worldwide basis including stores and offices. Apple now wants to include this for their suppliers.

CEO Tim Cook

Apple's commitment is impressive. According to CEO Tim Cook, it is out of the concern for the planet that we all share. As Cook puts it: "Businesses have a profound opportunity to help build a new sustainable environment" to sustain the planet we all share.

36.   New Tech Cuts BBQ Pollution 90%

Source: Stock Image of Barbecue

German Innovation

German scientists at the Fraunhofer Institute for Building Physics (IBP) have invented a technology that cuts toxic carbon fumes from barbecues by 90%. The IBP team studied and analyzed the toxic pollutants emitted by charcoal grills, pizza ovens and bread ovens and innovated a new system to greatly reduce it. Their research was commissioned by the German Environment Agency. By some estimates, the pollution emitted from a barbecue for a family of four is the equivalent of driving a car for 90 miles.

BBQ Pollution

The IBP scientists demonstrated that the BBQ grill produces nitrous oxide, carbon monoxide, particulate matter, benzene and hydrocarbons - many of which are harmful to human health.

They found that the pollution is produced by incomplete combustion of fats combining with the type of fuel used. The higher the fatty content of the meat, the higher the amount of pollution.

New Prototype Technology
The IBP scientists' solution is new technology that treats the fat before it generates toxic pollution. They say it reduces the exhaust fumes by 90%. Their technology is in the prototype stage with very promising commercial prospects.

37.   Flying on Hydrogen Fuel Cells

Source: ZeroAvia

Carbon Free, Sustainable Flying
ZeroAvia, headquartered in Hollister, CA, is building aircraft powered by hydrogen fuel cells. It recently flew a modified aircraft with a 350 horsepower powertrain, driven by hydrogen fuel cells. The US company is part of the UK's HyFlyer initiative which is targeted at moving electric aircraft away from batteries toward fuel cells.

Hydrogen Flights
ZeroAvia argues that hydrogen fuel cells are highly effective and efficient for electric planes. They are far lighter and less volatile than batteries. In fact the ignition point for hydrogen is much higher than for jet fuel or lithium-ion batteries. Because hydrogen is so light, it tends to disperse rather than collect. Most im-

portantly an electric powertrain powered by hydrogen fuel cells is twice as efficient as an internal combustion engine.

Hydrogen's Future

The current drag on the use of hydrogen fuel cells for electric planes is not the technology. It is the lack of regulatory standards, which needs to be addressed. In the meantime, ZeroAvi will have a 10 to 20 seat aircraft on the market by 2023. And, they believe 200 seat hydrogen aircraft travelling 3,000 miles will be operational within 20 years.

38. Solar Powered Trains in India

Source: Indian Railway

India's Green Experiment

This is unique green innovation from India. A large solar power plant has been built in Bina, India to directly feed solar powered energy to overhead power lines to power electric locomotives. This is a major project of the Piush Goyaled Indian Railways. The project is considered a world first.

Testing Phase

This is a joint project of Indian Railways and Bharat Heavy Electrical Limited. It's a 1.7 megawatt solar power plant. According to the companies, the plant can "produce 25 lakh of energy yearly and save an amount of Rs 1.37 crore for Indian Railways every year." They are testing the technology.

3 Gigawatt Solar Plant

The plan is to install a 3 gigawatt solar power plant within a few years, after completing tests on their 1.7 megawatt solar plant system. According to the Indian Railway Ministry, the power plant's feeding directly to the railway's overhead traction system involves highly innovative technology that converts Direct Current to simple phase Alternating Current. This green experiment is making way for the era of solar powered, electric locomotive trains.

39. Audi's Big Green Power Plans

Source: NASA

Green on Green Power

An environmentally friendly, innovative undertaking by German automaker Audi and utility EnBW to recycle and reuse old EV batteries is underway. They'll use the old batteries to help power the grid. By the end of 2020, they will have completed a new electric car battery operation at EmBW's Heilbrom plant to build scalable energy storage facilities. E-battery recycling and reuse are the holy grail for car and battery manufacturers. And the storage of energy is another big priority for industry.

Energy Recycling, Reuse, Storage

For renewable energy companies, storing excess, unused surplus energy from solar and wind power generation is an urgent need. So, Audi wants to develop what they're calling a "plug and play" battery storage system for the energy industry. Audi says this system will be a blueprint for a scalable product that they intend to go to market with.

Lots of Life in Old Batteries

According to Audi, a used EV battery can still run for another three to ten years. They will build energy storage containers at the new facility and then start selling them to industrial power companies, utilities and generator companies. This has the potential to be very needed, green, innovative energy technology on a global scale.

40.   Flying Carbon Free

Source: Concept Electric Plane

Greener Skies

The United Kingdom wants to be the first to develop a commercial jet plane to fly across the Atlantic without any carbon emissions. UK Transport Secretary Grant Shapps made the pledge. He was re-enforcing Prime Minister Boris Johnson promise that the UK would produce the world's first zero-emissions, long haul passenger plane.

JetZero
The British government has set up the JetZero Council to specifically move their program forward. A lot is entailed including sustainable aviation fuels, electric planes, hybrid planes and hydrogen planes.

Global Goals
There is a global goal to achieve net zero carbon emissions by 2050. That is a key goal that Prime Minister Johnson has in developing the world first zero emissions, long distance passenger planes.

41.   Climate Change Warning from Siberia

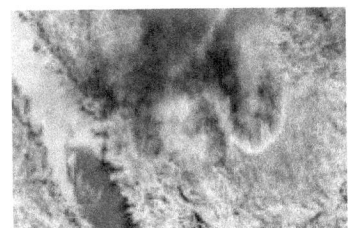

Source: Siberian Forest Fires in June 2020 via satellite image

Clear Impact of Climate Change Two Years in a Row
In June 2020, towns in northern Russia experienced sweltering heat normally found in the Tropics. In Siberia, it's was so hot wildfires were constantly breaking out, pine trees were bursting into flames and peat bogs were stone dry. In some locations, tem-

peratures hit 100 degrees (F). Climate scientists are calling the Siberian heat wave a "warning cry" about Climate Change from the Arctic. Experts at the World Meteorological Organization are troubled by satellite images showing that much of Russia's Arctic is in an extreme heat zone.

Blazing Heat

The extreme heat fanned forest fires and ignited normally water saturated peat bogs. Climate scientists fear that the intense heat and fires in Siberia are signs of even hotter and drier conditions to come with more frequent forest fires. Those fires release carbon stored in the forests and peatbogs, increasing the amount of greenhouse gases in the air that are warming the Earth.

Vicious Cycle and Pattern

This is a vicious pattern of heat occurring in the Russian Arctic that makes Climate Change worse. The numbers are staggering. In June 2019 and June 2020, the amount of emissions from Russian Arctic heat waves and forest fires were greater than for every June from 2003 through 2018. Sobering scientific facts on the accelerating impact of greenhouse gas emissions.

42.   Generator Harvests Energy from Shadows

Source: NUS Device

This is An Electrifying Invention

This new invention is called a shadow-effect energy generator. It's the stuff of science fiction but it's real. The device produces a current from the differences between bright light and shadowy areas. The bigger the contrast in light the more energy it produces. The device can produce enough energy to power a small electronic device like an electronic watch. The materials scientist inventor - Swee Ching Tan of the National University of Singapore - says this allows us to "harvest energy anyplace on Earth."

Shadows of Value
The new device is similar to a solar cell. Tan created the generator by placing a thin coat of gold over silicon. Light shining on the silicon energizes its electrons. With the gold layer, the generator produces a current when part of the device is in shadow.

Next Steps and Potential Skyscraper Future
The team is working to increase the amount of light the generators can absorb to boost their performance. In the future, these generators may produce enough energy from shadows around a solar array to power skyscrapers and homes.

43.   Green Sustainable Beer

A Toast to a Cleaner Environment
Beer, especially craft beers, are a growing, energy intensive industry. The number of US breweries grew 200% in the past decade
and production has increased 12% per year.

Saving the Planet Per 6-Pack
To cut greenhouse emissions and save energy, a growing number
of breweries are investing in sustainable production technologies. But will it pay off? Research by Indiana University says yes
it will. A majority of US beer drinkers are willing to pay the extra
cost. In fact, according to the University's survey, 59% are willing to pay extra. On average, sustainable beer production practices tack on an extra $1.30 per 6 pack.

Sustainable Innovation
Innovative, energy saving, sustainable practices being imple-
mented by a growing number of breweries include solar power,
onsite wastewater treatment plants, insulated brewery vessels
and recaptured steam from the brewing process. Beer is one of
the world's oldest and most popular drinks - in 3rd place behind
water and tea.

44.   Jordan Turning the Desert Green

Jordan's Sustainable Solar & Seawater Farm Project

Ambitious, Innovative Project

Jordon has embarked on a very ambitious and innovative project to turn the desert green for farming. They have a pilot project underway in a location that's .62 miles from the Israeli border and 9.3 miles from the Red Sea.

Feeding the World

One of the most pressing needs facing humanity is feeding the world's growing population. By 2050, food production needs to go up by 50% to feed the projected global population. The problem is further complicated by the lack of good water, which is a big issue in Jordan. Furthermore, food production totals 70% of global freshwater consumption and emits 25% of greenhouse gases. Changes are clearly needed. Jordan, which is the 2nd poorest water supply nation and three-quarters desert, is taking an innovative, problem solving lead.

Jordan Playing to Its Strengths

Jordan has plenty of sunshine and seawater. Their solar-seawater solution to making their desert green is both sustainable and simplicity itself. Jordan's solar energy desalinates the seawater, the desalinated water grows the crops, runoff cools greenhouses and the crops help plough back carbon into the soil.

Using Seawater to Farm Crops in the Desert

This innovative program is several years old. The seawater is being trucked in, but there are plans to build pipelines into the desert from the Red Sea. Plans are to scale up and commercialize this fascinating project and start making Jordan a sustainable source of native food for its population and also a food exporter.

45.   Geothermal Food Growing

Pagosa Springs Geodesic Greenhouses Powered by Geothermal Energy

A Grand Experiment: Geothermal, Renewable, Sustainable Farming

The small Colorado town of Pagosa Springs is using its hot springs to grow food for the community year round. It's a 5 year experiment by the Geothermal Greenhouse Partnership, a non-profit, volunteer organization that's pushing innovation with great potential.

Geothermal Energy 24/7/365

The group is turning the world's largest and deepest geothermal hot spring into a lot more than a tourist destination. They're using the renewable energy source that's available 24/7/365 to grow food year round for the community.

Geothermal Farming

Geothermal energy is energy harnessed from the earth's heat. Geothermal farming is something new, different and very promising. The reason is that geothermal energy is abundant in the US and could provide a renewable energy source for farming. The Colorado farming program could be a model for the US. Experts believe the potential for the US is huge.

Geodesic Greenhouses

The town has 3 geodesic greenhouses to house gardens. The program uses a closed loop water system where geothermal water is pumped in from town wells. It heats domestic water in the greenhouses and resumes its natural path. As a result almost no water is used. This is a program worth watching for green, sustainable, agriculture through renewable geothermal energy.

46. Discarded Plastic May Fuel Your Car

Source: Stock Photo

Plastics Pollution

Scientists at Swansea University in the UK say they've successfully converted discarded plastic into hydrogen fuel. The fuel can be used to power hydrogen powered cars and trains.

Billions of Tons of Plastics Dumped Every Year

Billions of tons of plastics are used every year and land in the dump and the oceans. Only a fraction of the plastic is recycled. The Swansea University researchers are trying to find a use for what's not being recycled.

Brilliantly Simple Idea and Process - Basis of Great Innovation

Here's the process. The scientists add light absorbing material to the plastic, before placing it in an alkaline solution and exposing it to light. Their process creates hydrogen. The UK scientists are quick to add that rolling out this breakthrough at an industrial level could take years. But wouldn't that be worth it!

47.   CityScapes 2050

Source: Stock Image of Future Buildings

Buildings Self-Sufficient & Sustainable Many futurists believe that because of climate change and depleted resources, future high rises and buildings in 2050 will have to be designed to be self-sufficient and sustainable. What would the key components of such buildings be? The vision of buildings and life in them in 2050 is fascinating.

Very Green Architecture
First of all, 2050 buildings will need to be energy self-sufficient. Renewable energy sources will be part of the architecture such as solar roofs and solar paneled windows. Algae biofuel cells will be part of the façade generating energy to power the buildings. Several floors will be dedicated to farming to provide the building's food supplies. Rain water will be collected and converted into drinking water. 3D, 4D and 5D printers will be ubiquitous to manufacture on the spot whatever is needed. Some futurists see the buildings as self-contained islands, where everything needed is created within.

2050: Life in Very Smart Cities

This is a fascinating, futuristic vision of life in 2050. Cityscapes of very smart buildings in very smart cities that are self-sufficient, self-sustaining and green.

48. Magic Carpet Ride

Scientists from California Institute of Technology and NASA's Jet Propulsion Lab are wishing on Aladdin's Lamp for a magic carpet ride. Their solar energy research project involves launching 2500 "magic carpets" tightly side by side into space orbit to absorb the sun's energy. The carpet system would beam the energy back to earth ground-stations to generate electricity.

Orbiting Carpets to Light the World

The carpets will fly over an area the size of 1,670 football fields or 3.5 square miles. The carpets are only one inch thick. They're designed to precisely beam energy back to earth to targeted locations. Primary targets would be global areas in need of energy, particularly in the third world. It would be the largest space structure ever built. And the targeted beam precision is provided by technology called "phased array" which is currently used in radar systems. The key is controlling the timing and patterns in which each solar tile turns its antenna on. Researchers say that enables the system to generate electromagnetic waves targeted to desired locations.

Space Solar Power Initiative
The project is formally called the Space Solar Power Initiative or SSPI. It's being led by three California Institute of Technology Professors who also serve as research scientists at NASA's Jet Propulsion Lab. SSPI is a partnership between Caltech and Northrup, which provided $17.5 million to fund development of key components.

Compelling Need
According to the United Nations, 25% of the world population doesn't have access to reliable electricity. Of that number, half have no electricity at all. The researchers believe their orbiting carpets are a solution. They compare their emerging technology to wireless phones. In parts of the third world, there are no telephone landlines. Wireless networks are easier and cheaper to deploy for phone service. Similarly in locations without power plants or transmission lines, it's easier to build ground stations and beam in energy from space. That energy would then be converted into electricity and transmitted to local communities. It's a magic carpet ride to light up the world.

49. Green Big Rigs

Hyundai's Prototype Hydrogen Fuel Cell Truck

Zero Emissions and Sustainable

Hyundai and Toyota are launching hydrogen fuel cell trucks. The trucks are green, no greenhouse gas emissions, sustainable and quiet.

Hyundai Plan
The South Korean automaker plans on deploying 1000 hydrogen fuel cell trucks in Switzerland in the early 2020's. Hyundai says the commercial trucks have an estimated range of 248 miles fueled by 8 onboard hydrogen storage tanks. Refueling takes 7 minutes.

Toyota Trucks on the Road
Toyota has deployed a big freight truck powered by hydrogen fuel cells. The truck has a range of 300 miles. Toyota has 3 partners in the project: the Port of Los Angeles, Shell and Kenworth. They're developing 10 zero emissions trucks. Toyota is providing fuel cell stacks and tanks, batteries and electric motors that will be put into Kenworth's Class 8 trucks. Toyota is also providing powertrains and operational support. The partners are setting up several hydrogen fueling stations.

50.   Whiskey Biofuel, Renewable and It Works

Source: Celtic Renewables - World's 1st Whiskey Powered Car

No Engine Modification Needed

It's called bio-butanol. It's a whiskey powered biofuel that's sustainable, renewable with no CO2 emissions. And, it works. Celtic Renewables, the startup behind the breakthrough biofuel, has demonstrated its viability by fueling a car with it and taking it for a spin. That was the world's 1st drive powered by whiskey biofuel. It's a case of Scotch whiskey powering Green transportation with no engine modification needed. The company calls it "next generation biofuel".

Scottish Ingredients for Spirited Biofuel
The biofuel was created in Edinburgh at Napier University by an Irish Professor Martin Tangney. His company Celtic Renewables works with a large distillery. The biofuel is produced from draft, the sugary barley kernels that make fermentation happen. Another ingredient is pot ale, the liquid loaded with yeast after the distillation of whiskey. It's an all-natural renewable energy source.

Whiskey for a Good Ecofriendly Purpose
Celtic Renewables is in the process of scaling up its business. They foresee a multibillion dollar global market to transport green on whiskey biofuel. Other countries are interested in this spirited biofuel including Japan.

51.   Supertanker Powered by Wind

Source: Maersk Six, Cylinder High Tech Sails

High Tech Sails

Oil prices are surging. Danish shipping giant Maersk is testing high tech, mechanical sails to cut its marine fuel costs. It's installed 100 foot tall rotating cylinders on one of its ocean-going vessels. The cylinders are very high tech sails designed to capture the power of the wind and convert it into energy.

Norsepower

Norsepower, based in Finland, makes the sails of composite material. They are rotor sails and are based on what's called the "Magnus Effect". In effect, a spinning object drags air faster around one side, creating a pressure difference that pushes the vessel in the direction of the lower pressure side.

Wind Testing Sustainable Energy

If the system proves successful, Maersk expects to start saving 10% in fuel bills per vessel. And it will install the technology on at least 80 of its 164 tanker fleet. It's green, sustainable energy that cuts emissions into the air and ocean.

52.   China's Super Intelligent Road

Source: China's Solar Expressway

Electric Cars Recharged by Smart Highways

The Chinese are leading the way to the future of transportation. The future promises autonomous vehicles on "intelligent" highways. The smart roads are paved with electric battery rechargers, mapping sensors and solar panels. The rechargers will repower electric cars as they drive along. Some call China's super intelligent highway "the Solar Expressway".

Tests Now Underway on the Solar Expressway

China's smart road has been built in the city of Jinan. It's 3,540 feet long and the technologies are being embedded under transparent concrete. The solar panels that are already embedded generate enough electricity to power highway lights and 800 nearby homes. 45,000 vehicles barrel down the smart highway daily.

Made in China 2025 Masterplan

This Chinese smart highway is the road to our transportation future. It's part of President Xi Jinping's plan to make China an advanced manufacturing, technology and innovation power by 2025. It's a smart, solar powered road to the future of smart cars. It costs about $6.5 million.

BOOKS BY JOURNALIST

EDWARD KANE

Non-fiction books on innovations and discoveries across industries in 2020 and 2019.

"BIG TRAVEL BREAKTHROUGHS 2020's"
Just published by Amazon and Kindle
https://amazon.com/author/ekane

TOP INVENTIONS FOR THE 2020's
Just published by Amazon and Kindle
amazon.com/author/ekane

SMART DEVICES FOR THE 2020's
Just published by Amazon and Kindle amazon.com/author/ekane

FUTURE TRAVEL VEHICLES

Just published by Amazon and Kindle amazon.com/author/ ekane

FUTURE TRAVEL VEHICLES BY EDWARD KANE

SPACE

- "Space 2020's: What's Up There?" - Kindle & paperback
- "Bargain Space Trips" - Kindle & paperback
- "Space Renaissance in the 21st Century" - Kindle & paperback
- "Search For Life in Space" - Kindle & paperback
- "Space: The Unknown Region" – Kindle & paperback

TRANSPORTATION AND TRAVEL INNOVATIONS

- "Future of Transportation: 2020's and Beyond" - Kindle & paperback
- "Electric Vehicles for All" - Kindle & paperback
- "Hot Electric Vehicles for the 2020's" - Kindle & paperback
- "Important Innovations: Transportation" - Kindle, paperback & Audiobook on Audible
- "How to Travel in the Future" Vol. 1 & 2 - Kindle & paperback

LISTS OF TOP NEW INNOVATIONS

- "List of Best New Innovations" - Kindle & paperback
- "List of Top New Environmental Innovations" - Kindle & paperback
- "List of Top New Gadgets" - Kindle & paperback
- "List of Top New Medical Innovations" - Kindle & paperback
- "List of Top New Energy Innovations" - Kindle, paperback & Audiobook on Audible
- "List of Top New Robots" - Kindle, paperback & Audiobook on Audible
- "How to Use AI & AR" - Kindle & paperback

INVESTING IN INNOVATIONS

- "Investing in Disruptive Innovations" - Kindle & paper-back

Fiction - Adventure, Life Lessons Book for Children

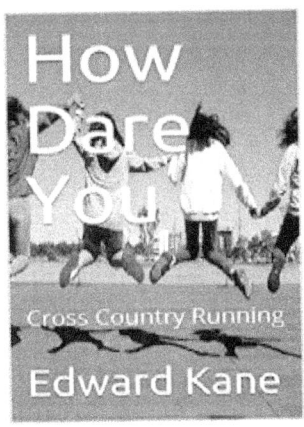

- "How Dare You" - Kindle, paperback & Audiobook on Audible

amazon.com/author/ekane

www.ingramcontent.com/pod-product-compliance
Lightning Source LLC
Chambersburg PA
CBHW070121230526
45472CB00004B/1356